D0971698

BASIC STATISTICS FOR BUSINESS AND ECONOMICS

Second Edition

I bought in Palm Springs,

The Wiley/Hamilton Series in
MANAGEMENT AND ADMINISTRATION

ELWOOD S. BUFFA, *Advisory Editor*
University of California, Los Angeles

PRINCIPLES OF MANAGEMENT: A MODERN APPROACH
FOURTH EDITION
Henry H. Albers

OPERATIONS MANAGEMENT: PROBLEMS AND MODELS
THIRD EDITION
Elwood S. Buffa

MODERN PRODUCTION MANAGEMENT: MANAGING THE OPERATIONS FUNCTION
FIFTH EDITION
Elwood S. Buffa

CASES IN OPERATIONS MANAGEMENT: A SYSTEMS APPROACH
James L. McKenney and Richard S. Rosenbloom

ORGANIZATIONS: STRUCTURE AND BEHAVIOR, VOLUME I
SECOND EDITION
Joseph A. Litterer

ORGANIZATIONS: SYSTEMS, CONTROL AND ADAPTATION, VOLUME II
Joseph A. Litterer

MANAGEMENT AND ORGANIZATIONAL BEHAVIOR: A MULTIDIMENSIONAL APPROACH
Billy J. Hodge and Herbert J. Johnson

MATHEMATICAL PROGRAMMING
SECOND EDITION
Claude McMillan

DECISION MAKING THROUGH OPERATIONS RESEARCH
SECOND EDITION
Robert J. Thierauf and Robert C. Klekamp

QUALITY CONTROL FOR MANAGERS & ENGINEERS
Elwood G. Kilpatrick

PRODUCTION SYSTEMS: PLANNING, ANALYSIS AND CONTROL
SECOND EDITION
James L. Riggs

COMPUTER SIMULATION OF HUMAN BEHAVIOR
John M. Dutton and William H. Starbuck

QUANTITATIVE BUSINESS ANALYSIS
David E. Smith

INTRODUCTION TO GAMING: MANAGEMENT DECISION SIMULATIONS
John G. H. Carlson and Michael J. Misshauk

PRINCIPLES OF MANAGEMENT AND ORGANIZATIONAL BEHAVIOR
Burt K. Scanlan

COMMUNICATION IN MODERN ORGANIZATIONS
George T. and Patricia B. Vardaman

THE ANALYSIS OF ORGANIZATIONS
SECOND EDITION
Joseph A. Litterer

COMPLEX MANAGERIAL DECISIONS INVOLVING MULTIPLE OBJECTIVES
Allan Easton

MANAGEMENT SYSTEMS
SECOND EDITION
Peter P. Schoderbek

ADMINISTRATIVE POLICY: TEXT AND CASES IN THE POLICY SCIENCES
Richard M. Hodgetts and Max S. Wortman, Jr.

THE ECONOMICS OF INTERNATIONAL BUSINESS
R. Hal Mason, Robert R. Miller and Dale R. Weigel

BASIC PRODUCTION MANAGEMENT
SECOND EDITION
Elwood S. Buffa

MANAGEMENT SCIENCE/ OPERATIONS RESEARCH:
Elwood S. Buffa and James S. Dyer

OPERATIONS MANAGEMENT: THE MANAGEMENT OF PRODUCTIVE SYSTEMS
Elwood S. Buffa

PERSONNEL ADMINISTRATION AND HUMAN RESOURCES MANAGEMENT
Andrew F. Sikula

MANAGEMENT PRINCIPLES AND PRACTICES
Robert J. Thierauf, Robert Klekamp and Daniel Geeding

BASIC STATISTICS FOR BUSINESS AND ECONOMICS

Second Edition

Paul G. Hoel and Raymond J. Jessen
UNIVERSITY OF CALIFORNIA, LOS ANGELES

A Wiley/Hamilton Publication
JOHN WILEY & SONS
New York • Chichester • Brisbane • Toronto

This book was designed by William
Tenney, copyedited by Susan Gerstein
and set in 10 point Plantin by
Applied Typographic Systems. The
cover was designed by Kathy Trainor
and printing and binding was done by
Halliday Lithographers. Chuck Pendergast
and Jean Varven supervised production.

Library of Congress Cataloging in Publication Data:

Hoel, Paul Gerhard, 1905-
Basic statistics for business and economics.

(The Wiley/Hamilton series in management and administration)
Includes index.
1. Statistics. I. Jessen, Raymond James, 1910- joint author. II. Title.
HA29.H66 1977 519.5 76-54504
ISBN 0-471-40268-0

Printed in the United States of America

10 9 8 7 6 5 4 3